D1446596

PowerPhonics™

Rain

Learning the AI Sound

Pam Vastola

The Rosen Publishing Group's
PowerKids Press™
New York

Rain falls from the clouds.

Rain falls on a flower.

Rain falls on Gail.

Rain falls on our dog's tail.

Rain falls on a snail.

Rain falls into a pail.

Rain falls on a train.

11.2092
NSB

Rain falls down the drain.

I wait for the rain to stop.

The rain falls drop by drop.

Word List

drain
Gail
pail
rain
snail
tail
train
wait

Instructional Guide

Note to Instructors:
One of the essential skills that enable a young child to read is the ability to associate letter-sound symbols and blend these sounds to form words. Phonics instruction can teach children a system that will help them decode unfamiliar words and, in turn, enhance their word-recognition skills. We offer a phonics-based series of books that are easy to read and understand. Each book pairs words and pictures that reinforce specific phonetic sounds in a logical sequence. Topics are based on curriculum goals appropriate for early readers in the areas of science, social studies, and health.

Letters/Sound: **ai** – Divide a chalkboard or dry-erase board into space for three columns. Write the words *make, train,* and *play* as headers for the columns. Have the child tell what vowel sound they hear in all three words. Have them underline the **long a** sound in each of the words.

- Have the child name additional words that have the **long a** sound. List the words in the three columns according to their spelling patterns.

Phonics Activities: Present the following words on flash cards: *train, pail, tail, Gail, mail, sail, wait, rain, nail.* Clarify meanings of unfamiliar words. Have them tell which word has five letters. Have them identify the word that is a girl's name.

- Write words with **ay** endings for the child to decode, such as: *pay, play, may, say, stay, gray, hay, lay,* etc. Have the child underline the **long a** in each word. Point out that **y** is silent in these words. Talk about how these words have the **long a** sound like the vocabulary words in *Rain,* but are spelled differently.
- For two or more children. Prepare playing cards with picture-words of mostly **ai** words. Include a few that have different long vowel sounds in them. Have two partners or small groups play "Go Fish" and try to collect the most pairs of words containing the **ai** sound. Ask the children to read the **ai** words to you.

Additional Resources:

- Branley, Franklyn M. *Down Comes the Rain.* New York: HarperCollins Publishers, 1997.
- Lynn, Sara, and Diane James. *Rain & Shine.* Chicago, IL: World Book, Inc., 1997.
- McKinney, Barbara S. *A Drop Around the World.* Nevada City, CA: Dawn Publications, 1998.

Published in 2002 by The Rosen Publishing Group, Inc.
29 East 21st Street, New York, NY 10010

Book Design: Ron A. Churley

Photo Credits: Cover © SW Productions/Index Stock; pp. 3, 21 © SuperStock; p. 5 © Ron Watts/Corbis; p. 7 © Benelux Press/Index Stock; p. 9 © Omni Photo Communications/Index Stock; p. 11 © Eric Sanford/International Stock; pp. 13, 17 by Karey L. Schuckers-Churley; p. 15 © Grayce Roessler/Index Stock; p. 19 © Patrick Ramsey/International Stock.

Library of Congress Cataloging-in-Publication Data

Vastola, Pam.
Rain : learning the AI sound / Pam Vastola.— 1st ed.
p. cm. — (Power phonics/phonics for the real world)
ISBN 0-8239-5943-0 (lib. bdg. : alk. paper)
ISBN 0-8239-8288-2 (pbk. : alk. paper)
6-pack ISBN 0-8239-9256-X
1. Rain and rainfall—Juvenile literature. 2. English language—Vowels—Juvenile literature. [1. Rain and rainfall.]
I. Title. II. Series.
QC924.7 .V37 2002
551.57'7—dc21

2001000381

Manufactured in the United States of America